Preface

Science Olympiad All Chapter Wise Story Book Class 2 is a thoughtfully crafted resource designed to make science learning an exciting adventure for young minds. Tailored specifically for Class 2 students, this book transforms scientific concepts into captivating stories, making them relatable, engaging, and easy to understand. Through vibrant narratives, colourful illustrations, and interactive activities, it inspires curiosity and fosters a love for science from an early age.

Aligned with the Science Olympiad syllabus, each chapter covers key topics such as plants, animals, the human body, earth, water, and air, ensuring a thorough understanding of the foundational concepts. The book goes beyond traditional learning by connecting scientific ideas to everyday life, making them memorable and practical. With hands-on exercises, playful questions, and thought-provoking activities, students are encouraged to think critically, solve problems, and explore the world around them.

Whether preparing for the Science Olympiad or simply exploring the wonders of science, this book is an invaluable tool for students, parents, and teachers. It not only equips students for academic success but also nurtures a lifelong passion for discovery and learning.

Table of Contents

Chapter 1 - The Invention of The Wheel

In the ancient village of Stoneville, people lived simple but hard lives. They had to carry heavy loads on their backs or drag them along the ground. Moving large stones to build houses or carrying food from the fields was backbreaking work. Life was tough, and everyone wished for a way to make things easier.

A young boy named Ravi lived in Stoneville. Ravi was curious and always loved to tinker with things. One day, while playing near the river, he saw round stones rolling down the hill. An idea sparked in his mind. "What if we could make something that rolls like these stones to carry heavy loads?" he thought.

Ravi ran home and shared his idea with his father, who was a carpenter. Together, they carved a large, smooth wooden circle, attached it to a wooden frame, and added another circle on the other side. They placed a basket on top, and to their amazement, it rolled easily across the ground. Ravi had invented the first wheel!

When Ravi showed his invention to the villagers, they were astonished. They called it the "Wheel of Change" because it made their lives so much easier. With the wheel, they could transport heavy stones, carry large amounts of food, and even travel longer distances. Stoneville was never the same again. The wheel became an essential part of their daily lives, leading the village to prosperity and growth.

The invention of the wheel taught the villagers that innovation and creativity could solve many of their problems. It was a turning point for Stoneville, bringing about positive change and progress.

Moral: Innovation and creativity have the power to transform lives and bring progress.

Activities:

1. Build a Simple Wheel Model:

Materials Needed: Cardboard, pencils, plastic straws, and tape.

Instructions: Cut out two circles from the cardboard to represent wheels. Attach them to either side of a pencil (the axle) using tape. Use a straw to help the pencil spin smoothly. Let the children roll their wheel models and observe how easily it moves.

2. Roll and Race:

Instructions: Bring small items like toy cars or rolling pins. Let the children race them down a slope to see how wheels help in moving objects faster.

Subjective Questions:

1. Why was life difficult for the people of Stoneville before the invention of the wheel?

2. What inspired Ravi to invent the wheel?

3. How did Ravi and his father create the first wheel?

4. How did the invention of the wheel change life in Stoneville?

5. Why do you think the villagers called the invention the "Wheel of Change"?

6. What are some ways we use wheels in our daily lives today?

7. How can creativity help solve problems?

8. Why is the wheel considered one of the most important inventions in history?

9. How did the villagers react to Ravi's invention?

10. What do you think Ravi felt after successfully inventing the wheel?

Chapter 2 - Good Habits

In the heart of Habitsville, there was a beautiful garden called the "Garden of Good Habits." The garden was magical because every time someone practiced a good habit, a new flower would bloom. The garden was tended by an elderly woman named Mrs. Bloom, who had a gentle smile and a kind heart.

A young boy named Rohan moved to Habitsville with his family. Rohan was not used to practicing good habits; he didn't brush his teeth regularly, forgot to wash his hands, and rarely said "please" or "thank you." His parents were worried, so they decided to take him to Mrs. Bloom.

Mrs. Bloom welcomed Rohan into the garden and explained how it worked. She said, "Every time you practice a good habit, you'll see a flower bloom in this garden. The more good habits you practice, the more beautiful the garden becomes."

Rohan decided to give it a try. The next morning, he brushed his teeth and greeted his parents with a smile. When he visited the garden later that day, a beautiful rose had bloomed. Rohan was amazed and excited. He began practicing more good habits—washing his hands before meals, helping his parents, and being polite to everyone he met. Each time he did something good, the garden bloomed with even more flowers.

Soon, Rohan's life changed. He felt happier and more confident. The garden became a symbol of his growth, and he loved visiting it to see the results of his good habits. Other children in the village started following Rohan's example, and soon, the Garden of Good Habits was the most beautiful place in Habitsville.

Moral: Good habits lead to personal growth and make the world around us more beautiful.

Activities:

1. Create a Good Habits Garden:

Materials Needed: Construction paper, markers, and scissors.

Instructions: Have the children create a paper garden by drawing flowers. Each time they practice a good habit, they can add a new flower to their garden.

2. Good Habits Chart:

Materials Needed: Paper, markers, and stickers.

Instructions: Help the children create a weekly chart where they can track their good habits (e.g., brushing teeth, saying thank you). They can add a sticker each time they complete a good habit.

Subjective Questions:

1. What made the Garden of Good Habits special?

2. How did Rohan feel about practicing good habits at first?

3. What happened when Rohan started practicing good habits?

4. How did the garden change as Rohan practiced more good habits?

5. Why did other children in the village start following Rohan's example?

6. What are some good habits you practice at home or school?

7. How do good habits make life better for you and those around you?

8. Why is it important to be consistent in practicing good habits?

9. How can you encourage others to practice good habits?

10. What did Rohan learn from his experience in the Garden of Good Habits?

Chapter 3 - Means of Communication

In the ancient kingdom of Communica, there were two towns, Talksville and Writesburg. The people of Talksville loved to speak to each other face-to-face, while the people of Writesburg preferred to write letters. Although the towns were close, there was a deep river that separated them, and they rarely communicated with each other.

One day, a wise old woman named Lia visited both towns. She noticed that Talksville had many ideas but lacked ways to preserve them, while Writesburg had beautifully written letters but missed the warmth of spoken words. Lia had an idea to bring the towns together.

She proposed building two bridges across the river. One bridge would be called "The Bridge of Words" where people could speak to each other, and the other would be "The Bridge of Letters" for sending written messages. The towns agreed and began construction.

Once the bridges were complete, people from Talksville visited Writesburg to exchange ideas in person, while people from Writesburg sent beautifully crafted letters to Talksville. The bridges made communication easier, and soon, the two towns grew closer and shared their cultures and knowledge. The kingdom of Communica flourished, and everyone realized that both speaking and writing were important ways to connect with others.

Moral: Different forms of communication help bridge gaps and bring people closer together.

Activities:

1. Write and Deliver a Message:

Materials Needed: Paper, pencils, envelopes.

Instructions: Have the children write a short letter or message to a friend or family member. Encourage them to deliver the message in person or send it through the mail.

2. Communication Role-Play:

Instructions: Divide the children into pairs. One child will act as a resident of Talksville (speaking their message), and the other will act as a resident of Writesburg (writing their message). Discuss how different forms of communication can be effective.

Subjective Questions:

1. What were the names of the two towns in the kingdom of Communica?

2. Why did Talksville and Writesburg rarely communicate with each other?

3. What did Lia propose to help the towns communicate better?

4. How did the bridges change the relationship between Talksville and Writesburg?

5. What did the people of Talksville learn from the people of Writesburg, and vice versa?

6. Why is it important to have different forms of communication?

7. How can communication help bring people closer together?

8. What is your favorite way to communicate with your friends and family?

9. How do you think life would be different without the ability to communicate?

10. What did the kingdom of Communica learn from building the bridges?

Chapter 4 - Clothes

In the vibrant town of Weaveville, everyone wore clothes that represented their heritage and traditions. The town was known for its beautiful tapestries, which were woven from different fabrics brought by travelers from around the world. Each tapestry told a story of the people who lived there.

A young girl named Mira was fascinated by the tapestries and the stories they told. She often visited the town's weaving hall, where the elder weavers would work tirelessly to create the intricate designs. One day, Mira noticed that the weavers were arguing about which fabric was the most important—some believed that silk was the finest, others argued for cotton, and some preferred wool.

Mira was saddened by the argument and decided to find a way to bring everyone together. She traveled to the neighboring towns, where she learned about the different types of fabrics and the significance they held for each culture. In Silkville, she discovered how silk represented elegance and was used in special ceremonies. In Cottonland, she learned how cotton was valued for its comfort and durability, especially in hot climates. In Woolshire, she saw how wool kept people warm during cold winters and was a symbol of resilience.

Mira returned to Weaveville with stories and samples from each town. She gathered the weavers and shared what she had learned, explaining that each fabric was important in its own way and that the beauty of their town's tapestries came from the combination of all these different materials.

The weavers realized that they had been too focused on their own preferences and had forgotten the true purpose of the tapestry—to celebrate the diversity and unity of their community. They decided to work together, using silk, cotton, and wool to create a new tapestry that would symbolize the harmony of Weaveville.

The new tapestry was the most beautiful one they had ever made. It told the story of how different cultures and traditions could come together to create something truly special. The tapestry became a symbol of the town's unity, and Mira was celebrated for her wisdom and understanding.

Moral: Embracing and respecting diversity can lead to greater unity and beauty.

Activities:

1. Fabric Collage:

Materials Needed: Scraps of different fabrics (silk, cotton, wool, etc.), glue, and paper.

Instructions: Have the children create a collage using different fabric scraps to represent the diversity of materials. Discuss the significance of each fabric and how they all contribute to the final piece.

2. Dress-Up Day:

Instructions: Organize a day where children can come dressed in clothes that represent different cultures or their own heritage. Let them share the story behind their outfit and what it means to them.

Subjective Questions:

1. What made the town of Weaveville special?

2. Why did the weavers in Weaveville start arguing?

3. What did Mira learn from her travels to neighboring towns?

4. How did Mira help the weavers understand the importance of all the fabrics?

5. What was the significance of the new tapestry created by the weavers?

6. Why is it important to celebrate diversity in a community?

7. How do different fabrics represent different cultures and traditions?

8. What lesson did the weavers learn about working together?

9. How can we show respect for other cultures in our daily lives?

10. What did the tapestry symbolize for the town of Weaveville?

Chapter 5 - World Around Me

In the peaceful village of Naturia, there was a young boy named Arjun who loved exploring the world around him. Every day, Arjun would venture out to discover new things about nature whether it was the birds singing in the trees, the fish swimming in the clear rivers, or the flowers blooming in the meadows.

One day, Arjun found an old, tattered book in his grandfather's attic. The book was titled "The Journey of Discovery" and it had a map that led to hidden treasures in the nearby forest. Excited by the thought of an adventure, Arjun packed his backpack and set off into the forest, following the map.

As Arjun walked deeper into the forest, he noticed the different types of plants and animals he had never seen before. He saw how the sunlight filtered through the leaves, creating a magical glow on the forest floor. He heard the rustling of leaves as small animals scurried around, and the chirping of birds that filled the air with music.

The map led Arjun to a series of challenges. He had to cross a narrow bridge over a river, climb a steep hill, and find his way through a maze of trees. But with each challenge, Arjun discovered something new—he learned how to recognize different bird calls, how to identify edible berries, and how to find his way using the position of the sun.

Finally, Arjun reached the spot marked "X" on the map. Instead of finding a chest filled with gold, he found a simple wooden box. Inside the box was a note that read, "The true treasure is the knowledge and appreciation of the world around you. Cherish it, protect it, and share it with others."

Arjun realized that the journey itself was the real treasure. He had gained a deeper understanding and appreciation of nature, and he felt a strong connection to the world around him. When he returned home, he shared his experiences with his family and friends, inspiring them to explore and appreciate the beauty of nature as well.

Moral: The journey of discovering and appreciating the natural world is the greatest treasure.

Activities:

1. Nature Scavenger Hunt:

Materials Needed: A list of natural items to find (e.g., a leaf, a feather, a flower), a bag, and a magnifying glass.

Instructions: Take the children on a nature walk and give them a list of items to find. Encourage them to observe the details of each item and discuss its significance in nature.

2. Create a Nature Journal:

Materials Needed: A notebook, crayons, pencils, and stickers.

Instructions: Encourage the children to keep a nature journal where they can draw or write about the things they see in nature. They can include leaves, flowers, or small rocks they find and note how these items make them feel or why they are important.

Subjective Questions:

1. What did Arjun find in his grandfather's attic, and where did it lead him?

2. Describe some of the things Arjun saw during his journey through the forest.

3. How did Arjun overcome the challenges he faced on his journey?

4. What was the treasure that Arjun found at the end of his journey?

5. Why is it important to appreciate and protect the natural world?

6. What did Arjun learn about nature during his adventure?

7. How can exploring nature help us learn new things?

8. What can you do to protect the environment around you?

9. How did Arjun's journey inspire others in his village?

10. Why do you think the journey was more valuable than finding a chest of gold?

Chapter 6 - Get To Know Me

In the heart of the peaceful town of Harmonyville, there stood a grand old tree known as the Friendship Tree. This tree was special because it was believed that anyone who sat under its branches would form a lasting bond of friendship with the person next to them.

A new girl named Aisha had just moved to Harmonyville with her family. She was shy and didn't know anyone in the town. One day, her parents suggested that she visit the Friendship Tree to make new friends. Aisha was hesitant at first, but she decided to give it a try.

When Aisha arrived at the Friendship Tree, she found a boy named Sameer sitting alone, reading a book. She gathered her courage and sat down next to him. The tree's branches rustled as if welcoming her. Aisha and Sameer started talking, and they quickly discovered that they both loved books and had many other things in common.

As the days went by, Aisha and Sameer visited the Friendship Tree regularly, each time learning more about each other. They shared their hopes, dreams, and even their fears. They soon realized that their friendship had grown strong and deep, just like the roots of the tree.

The Friendship Tree became their favorite spot, where they could be themselves and feel understood. They often invited other children to join them under the tree, and soon, the Friendship Tree became a gathering place for all the children in Harmonyville.

Aisha learned that getting to know someone took time, patience, and a willingness to listen. But the rewards were worth it—she had found a true friend in Sameer, and together, they had created a place where everyone felt welcome.

Moral: True friendship is built on understanding, trust, and the willingness to get to know each other.

Activities:

1. Friendship Bracelets:

Materials Needed: Colorful threads, beads, and scissors.

Instructions: Have the children make friendship bracelets for each other. As they work, encourage them to talk about their interests and what makes them a good friend.

2. Friendship Circle:

Instructions: Have the children sit in a circle and take turns sharing something about themselves what they like, their favorite activities, or a fun fact. This activity helps them get to know each other better and build a sense of community.

Subjective Questions:

1. What made the Friendship Tree special in Harmonyville?

2. Why did Aisha feel hesitant to visit the Friendship Tree at first?

3. How did Aisha and Sameer become friends?

4. What did Aisha and Sameer discover they had in common?

5. Why is it important to take the time to get to know someone?

6. How can understanding and trust strengthen a friendship?

7. What role did the Friendship Tree play in bringing the children of Harmonyville together?

8. How can you be a good friend to others?

9. What did Aisha learn about friendship through her experience?

10. Why do you think the Friendship Tree was a symbol of lasting friendship?

Chapter 7 - Understanding Directions

In the mystical land of Naviland, everyone had a special compass that helped them find their way through life. These compasses didn't just point north; they guided people towards making the right choices and finding their true path.

A young boy named Arjun lived in Naviland and often got lost, not just in the forest, but also in making decisions in his everyday life. He found it hard to choose the right friends, decide what to do when faced with a challenge, and even figure out what he wanted to be when he grew up. Arjun often felt confused and unsure of himself.

One day, Arjun's grandfather, who was a wise and respected man in Naviland, gave him an ancient compass that had been passed down through generations. "This compass," his grandfather said, "is not just an ordinary one. It will guide you, not only in finding your way on a map but also in making the right decisions in life."

Curious and hopeful, Arjun decided to test the compass. The next day, he ventured into the deep forest with his compass in hand. As he walked, he came across a fork in the road. One path was well-trodden and easy, while the other was narrow and looked more difficult. Arjun wasn't sure which path to take. He took out the compass, and to his surprise, it didn't just point north. It gently pointed towards the narrow path.

Trusting the compass, Arjun chose the narrow path. Along the way, he faced challenges a steep hill to climb, a dark cave to pass through, and a fast-flowing river to cross. But each time, the compass guided him, showing him the safest way to overcome these obstacles. Finally, the path led Arjun to a beautiful valley filled with flowers and a clear stream, a place of peace and beauty that he had never seen before.

Arjun realized that the compass had not only guided him through the forest but also taught him valuable lessons about life. It showed him that the right path is not always the easiest one, but it is the one that leads to the most rewarding destinations. The compass helped him understand that making good decisions requires courage, patience, and trust in oneself.

As Arjun grew older, he continued to use the compass, both literally and metaphorically. Whenever he was faced with a difficult decision, he would think of the compass and choose the path that felt right in his heart, even if it was not the easiest choice. The compass became a symbol of his journey towards wisdom and self-discovery.

Moral: The right path in life is not always the easiest one, but it is the path that leads to true fulfillment and growth.

Activities:

1. Create a Simple Compass:

Materials Needed: A small magnet, a needle, a piece of cork, and a bowl of water.

Instructions: Show the children how to magnetize a needle by rubbing it against a magnet. Then, carefully stick the needle through a piece of cork and float it in a bowl of water. Watch as the needle points north. Discuss how a compass helps us find directions and how it can be a metaphor for making good choices in life.

2. Direction Hunt:

Instructions: Set up a simple treasure hunt around the classroom or playground where children have to follow directions (north, south, east, west) to find hidden objects. Discuss how following directions leads us to our goals and how it's similar to making decisions in life.

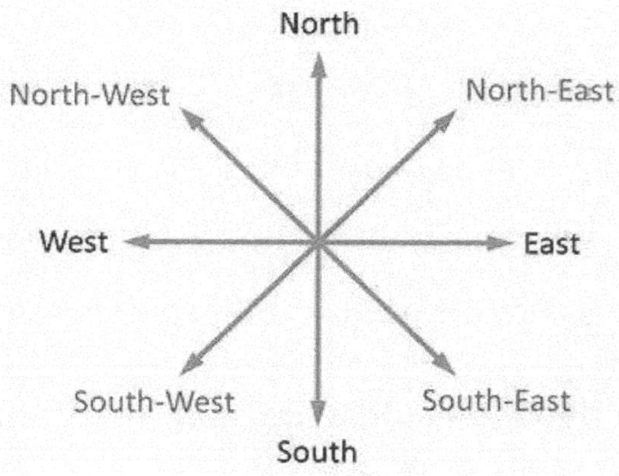

Subjective Questions:

1. What was special about the compass Arjun's grandfather gave him?

2. How did the compass help Arjun in the forest?

3. Why did Arjun choose the narrow path, and what did he learn from it?

4. What challenges did Arjun face on the narrow path, and how did the compass guide him through them?

5. What did Arjun discover at the end of his journey in the forest?

6. How can the compass be a metaphor for making decisions in life?

7. Why is it important to sometimes choose the harder path in life?

8. How did the compass help Arjun grow wiser and more confident?

9. Can you think of a time when you had to make a difficult decision? How did you find the right path?

10. What lessons did Arjun learn about life from his journey with the compass?

Chapter 8 - Houses

In the creative town of Buildsville, every family had a unique house that reflected their dreams, personalities, and values. The town was famous for its diversity in architecture—there were tall, sturdy brick houses, cozy wooden cottages, colorful clay homes, and even houses made of glass that sparkled in the sunlight.

A little girl named Anya lived in Buildsville, and she dreamed of building her own house one day. She spent hours imagining what it would look like. Would it be tall and majestic, or small and cozy? Would it have bright colors or be made of natural materials? Anya couldn't decide, so she decided to visit the different houses in Buildsville to find inspiration.

Her first stop was the brick house of Mr. Strong, who was a builder. Mr. Strong's house was tall and firm, with strong walls that could withstand any storm. He told Anya, "My house is built to last, just like my determination. I wanted a home that would protect my family and stand the test of time."

Next, Anya visited the wooden cottage of Mrs. Warm, a teacher who loved nature. Mrs. Warm's house was surrounded by trees and flowers, and the inside was filled with wooden furniture and cozy blankets. "I built my house from wood because I love the warmth of natural materials. It makes me feel close to the earth," she explained.

Anya then visited the colorful clay house of Mr. Joy, an artist. His house was painted in bright reds, blues, and yellows, with intricate designs on the walls. "I wanted my house to be a reflection of my creativity and happiness. Every color and pattern here tells a story," Mr. Joy said with a smile.

Finally, Anya visited the glass house of Ms. Clear, a scientist. The glass walls of her house allowed sunlight to pour in, creating a bright and open space. "My house is transparent because I believe in clarity and honesty. I want to see the world around me and let the world see who I am," Ms. Clear explained.

After visiting all these houses, Anya realized that each house was a reflection of the person who lived there. She understood that her dream house should reflect her own personality and values. She decided to build a house that was both strong and cozy, with a mix of bright colors and natural materials—a place where she could feel safe, happy, and inspired.

With the help of her family and friends, Anya built her dream house in Buildsville. It became a place where she could express herself, feel connected to nature, and welcome everyone who visited with warmth and joy.

Moral: Our homes reflect who we are and what we value; they are spaces where we express our dreams and find comfort and inspiration.

Activities:

1. Design Your Dream House:

Materials Needed: Paper, pencils, crayons, and markers.

Instructions: Have the children draw and design their dream house. Encourage them to think about what materials they would use, what colors they would choose, and how their house would reflect their personality.

2. Build a Miniature House:

Materials Needed: Cardboard, glue, scissors, and various decorating materials (e.g., fabric scraps, paint, etc.).

Instructions: Guide the children in building a small model of a house using cardboard and other materials. Discuss what each part of the house represents and how it reflects the builder's personality and dreams.

Subjective Questions:

1. What made the houses in Buildsville unique?

2. How did Mr. Strong's brick house reflect his personality?

3. Why did Mrs. Warm choose to build her house from wood?

4. What did Mr. Joy's colorful clay house represent?

5. How did Ms. Clear's glass house reflect her values as a scientist?

6. What did Anya learn from visiting the different houses in Buildsville?

7. How did Anya decide what her dream house would look like?

8. Why is it important for a home to reflect the personality and values of the people who live there?

9. What kind of house would you build to reflect your personality? Describe it.

10. How can our homes be a source of comfort, inspiration, and self-expression?

Chapter 9 - Services In The Neighborhood

Harmony Street was the most peaceful and friendly street in the town of Careville. The people who lived on Harmony Street cared deeply for each other and always worked together to keep their neighborhood safe and happy. Each person on the street had a special role to play, providing services that helped everyone.

At the corner of Harmony Street lived Mr. Fixit, the handyman. Whenever someone's door creaked, a window was stuck, or a faucet dripped, Mr. Fixit was there with his toolbox, ready to repair anything. He believed that a well-maintained home was the key to a comfortable life.

Next to Mr. Fixit lived Nurse Nina, who worked at the local clinic. She was always ready with a smile and a bandage for scraped knees or advice for staying healthy. She often visited her neighbors to check on their health, especially the elderly, and taught children about the importance of hygiene and cleanliness.

Further down the street was Ms. Green, the gardener. She had the most beautiful garden on Harmony Street, filled with flowers, vegetables, and herbs. Ms. Green believed that a green neighborhood was a happy one, so she shared her plants and gardening tips with everyone. She even organized community gardening days where all the neighbors would come together to plant trees and flowers, making Harmony Street even more beautiful.

At the end of Harmony Street lived Officer Safe, the neighborhood police officer. He made sure that everyone followed the rules the forest, but also in making decisions in his everyday life. He found it hard to choose the right friends, decide what to do when faced with a challenge, and even figure out what he wanted to be when he grew up. Arjun often felt confused and unsure of himself.

One day, Arjun's grandfather, who was a wise and respected man in Naviland, gave him an ancient compass that had been passed down through generations. "This compass," his grandfather said, "is not just an ordinary one. It will guide you, not only in finding your way on a map but also in making the right decisions in life."

Curious and hopeful, Arjun decided to test the compass. The next day, he ventured into the deep forest with his compass in hand. As he walked, he came across a fork in the road. One path was well-trodden and easy, while the other was narrow and looked more difficult. Arjun wasn't sure which path to take. He took out the compass, and to his surprise, it didn't just point north. It gently pointed towards the narrow path.

Trusting the compass, Arjun chose the narrow path. Along the way, he faced challenges—a steep hill to climb, a dark cave to pass through, and a fast-flowing river to cross. But each time, the compass guided him, showing him the safest way to overcome these obstacles. Finally, the path led Arjun to a beautiful valley filled with flowers and a clear stream, a place of peace and beauty that he had never seen before.

Arjun realized that the compass had not only guided him through the forest but also taught him valuable lessons about life. It showed him that the right path is not always the easiest one, but it is the one that leads to the most rewarding destinations. The compass helped him understand that making good decisions requires courage, patience, and trust in oneself.

As Arjun grew older, he continued to use the compass, both literally and metaphorically. Whenever he was faced with a difficult decision, he would think of the compass and choose the path that felt right in his heart, even if it was not the easiest choice. The compass became a symbol of his journey towards wisdom and self-discovery.

Moral: The right path in life is not always the easiest one, but it is the path that leads to true fulfillment and growth.

Activities:

1. Create a Simple Compass:

Materials Needed: A small magnet, a needle, a piece of cork, and a bowl of water.

Instructions: Show the children how to magnetize a needle by rubbing it against a magnet. Then, carefully stick the needle through a piece of cork and float it in a bowl of water. Watch as the needle points north. Discuss how a compass helps us find directions and how it can be a metaphor for making good choices in life.

2. Direction Hunt:

Instructions: Set up a simple treasure hunt around the classroom or playground where children have to follow directions (north, south, east, west) to find hidden objects. Discuss how following directions leads us to our goals and how it's similar to making decisions in life.

Subjective Questions:

1. What was special about the compass Arjun's grandfather gave him?

2. How did the compass help Arjun in the forest?

3. Why did Arjun choose the narrow path, and what did he learn from it?

4. What challenges did Arjun face on the narrow path, and how did the compass guide him through them?

5. What did Arjun discover at the end of his journey in the forest?

6. How can the compass be a metaphor for making decisions in life?

7. Why is it important to sometimes choose the harder path in life?

8. How did the compass help Arjun grow wiser and more confident?

9. Can you think of a time when you had to make a difficult decision? How did you find the right path?

10. What lessons did Arjun learn about life from his journey with the compass?

Chapter 10 - Time

In the bustling town of Tick-Tock, there lived a wise old clockmaker named Mr. Chronos. Mr. Chronos was known for creating the most accurate and beautiful clocks in the town. His shop was filled with all kinds of clocks—grandfather clocks, wall clocks, pocket watches, and even tiny wristwatches. People from all over Tick-Tock would come to him to have their clocks repaired or to buy a new one.

But Mr. Chronos was not just a clockmaker; he was also a philosopher. He often said that time was the most precious resource anyone could have. “Time, once lost, can never be regained,” he would remind his customers. “Use it wisely.”

One day, a young boy named Leo came to Mr. Chronos’s shop. Leo was always late for school, late for meals, and late for everything. He didn’t understand the importance of time and often wasted it playing games or daydreaming. His parents were worried, so they brought him to Mr. Chronos, hoping the old man could teach him a lesson.

Mr. Chronos welcomed Leo into his shop and showed him around. He pointed to a large hourglass filled with sand slowly trickling from one side to the other. “This hourglass,” Mr. Chronos explained, “represents your life. The sand represents time. Once the sand has fallen, you can’t turn the hourglass back to get that time again.”

Leo watched the sand fall and realized how quickly time was passing. Mr. Chronos then gave him a pocket watch and said, “This watch is a gift. It will remind you that time is always ticking. You must decide how to use it.”

Leo took the watch and began to pay more attention to how he spent his time. He started waking up early, arriving at school on time, and finishing his homework before playing. He realized that by managing his time better, he could enjoy both his responsibilities and his free time without feeling rushed or stressed.

As the weeks passed, Leo's life changed for the better. He felt more organized and happy. He thanked Mr. Chronos for teaching him the value of time and promised to use it wisely for the rest of his life.

Moral: Time is a precious resource that should be used wisely and thoughtfully.

Activities:

1. Make Your Own Clock:

Materials Needed: Paper plates, markers, clock hands cut out of cardboard, and a paper fastener.

Instructions: Have the children create their own clocks using the materials provided. Teach them how to tell time and discuss the importance of managing their time wisely.

2. Time Management Chart:

Materials Needed: Paper, rulers, markers, and stickers.

Instructions: Help the children create a daily time management chart, where they can plan their activities, including time for school, homework, play, and rest. Discuss how planning their time can help them accomplish more without feeling rushed.

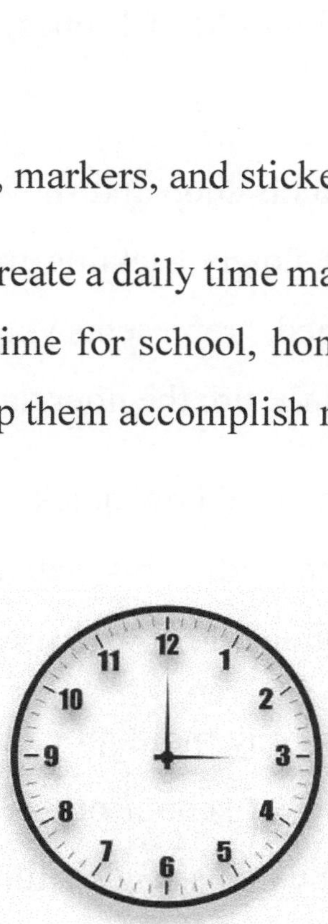

Subjective Questions:

1. What was Mr. Chronos known for in the town of Tick-Tock?

2. How did Mr. Chronos explain the concept of time to Leo?

3. What did the hourglass in Mr. Chronos's shop represent?

4. How did the pocket watch help Leo manage his time better?

5. What changes did Leo make in his life after learning about the importance of time?

6. Why is it important to manage your time wisely?

7. How can planning your day help you achieve more?

8. What are some activities that you need to make time for every day?

9. How can wasting time affect your responsibilities and your free time?

10. What lesson did Leo learn about time, and how can you apply it to your own life?

Chapter 11 - Air, Water, Food, Shelter

In the town of Essentia, the people lived in harmony with nature, understanding the importance of the four essentials of life: air, water, food, and shelter. These four elements were respected and protected by everyone, as they knew that without them, life would not be possible.

The town was surrounded by lush forests, clear rivers, fertile fields, and strong, sturdy homes. The air was fresh and clean, the water pure and abundant, the food grown with care, and the shelters built to withstand any weather. Essentia was a place where people thrived because they understood the balance between nature and human life.

One day, a group of travelers arrived in Essentia. They were amazed by how healthy and happy the people were. The travelers asked the townspeople how they had created such a perfect place to live. The elders of Essentia gathered the travelers and told them the story of the Four Guardians.

Long ago, Essentia was just a barren land, with no trees, rivers, or homes. But four wise guardians—Wind, Water, Earth, and Stone—came together to create a place where life could flourish. Wind provided the air, Water filled the rivers and streams, Earth grew the food, and Stone built the shelters.

The guardians taught the people of Essentia to respect these elements. Wind told them to keep the air clean by not polluting it. Water reminded them to conserve water and keep it pure. Earth showed them how to grow food sustainably, and Stone taught them to build strong homes that would protect them from the elements.

The people of Essentia followed the guardians' teachings and worked together to create a thriving community. They planted trees to purify the air, built wells to ensure a steady supply of water, cultivated crops to feed everyone, and constructed homes that were safe and comfortable.

The travelers were inspired by the story of the Four Guardians and decided to take these lessons back to their own towns. They realized that by respecting and protecting the four essentials of life, they could create a better world for themselves and future generations.

Moral: The four essentials of life—air, water, food, and shelter—must be respected and protected for life to thrive.

Activities:

1. Four Essentials Collage:

Materials Needed: Old magazines, scissors, glue, and paper.

Instructions: Have the children create a collage representing air, water, food, and shelter. They can cut out pictures from magazines or draw their own illustrations to show the importance of each essential element.

2. Nature Walk:

Instructions: Take the children on a nature walk to explore the four essentials in their environment. Discuss how trees provide oxygen, how water sources are important, how food is grown, and how homes protect us. Encourage them to think about ways to protect these essentials.

Subjective Questions:

1. What are the four essentials of life, according to the story?

2. Who were the Four Guardians, and what did each one represent in the town of Essentia?

3. How did the people of Essentia learn to respect and protect the four essentials?

4. What lessons did Wind teach the people about air?

5. How did Water encourage the people to take care of their water supply?

6. What sustainable practices did Earth show the people for growing food?

7. How did Stone help the people build strong and safe shelters?

8. Why is it important to protect the air, water, food, and shelter in our communities?

9. How did the story of the Four Guardians inspire the travelers who visited Essentia?

10. What can you do in your daily life to help protect the four essentials of life?

Chapter 12 - Means Of Transport

In the bustling city of Transitaria, people were always on the move. The city was known for its advanced means of transport, from bicycles and cars to trains, ships, and even airplanes. Every citizen of Transitaria had a favorite way to travel, and the city was always filled with the hum of engines, the clatter of train tracks, and the whoosh of airplanes taking off.

A young boy named Sam lived in Transitaria. Sam was fascinated by all the different ways people could travel. He loved riding his bicycle to school, watching the cars zoom by on the highway, and hearing the distant whistle of the trains. But what Sam really dreamed of was flying in an airplane.

One day, Sam's father, who was a pilot, decided to surprise him. "Sam," he said, "how would you like to come with me on a special journey to see the world from the sky?" Sam was overjoyed. He had always wondered what it would be like to fly above the clouds and see the world from a bird's-eye view.

As they boarded the airplane, Sam's father explained how each mode of transport had its own unique purpose. "Bicycles are great for short trips and staying healthy," he said. "Cars and buses help us travel longer distances on the ground, trains can take us across countries, ships carry us across oceans, and airplanes let us reach faraway places quickly."

As the plane took off, Sam looked out of the window in awe. The city below grew smaller and smaller until it looked like a tiny model. He saw the winding roads where cars were traveling, the railways where trains were speeding along, and even the ships in the harbor. From high above, he realized how all the different modes of transport worked together to connect people and places.

Their journey took them to new cities and countries, each with its own unique transportation systems. Sam learned about bullet trains in Japan, gondolas in Venice, and even hot air balloons in Turkey. He marveled at how transport not only made life easier but also allowed people to explore the world and experience new cultures.

When they returned home, Sam had a newfound appreciation for all the ways people traveled. He realized that each mode of transport had its own special role in making the world accessible and connected. Sam decided that when he grew up, he wanted to help design new and better ways for people to travel, so everyone could explore the world just like he did.

Moral: Each mode of transport plays an important role in connecting people and places, making the world accessible and bringing us closer together.

Activities:

1. Transport Collage:

Materials Needed: Old magazines, scissors, glue, and paper.

Instructions: Have the children create a collage of different modes of transport. They can cut out pictures of bicycles, cars, trains, ships, and airplanes, and arrange them on a poster to show how they connect people and places.

2. Build a Paper Airplane:

Materials Needed: Paper, markers, and stickers.

Instructions: Teach the children how to fold a paper airplane. Let them decorate their airplanes and then have a fun flying competition. Discuss how airplanes allow us to travel long distances quickly.

Subjective Questions:

1. What was Sam's favorite dream about transportation?

2. How did Sam's father surprise him, and what did they do together?

3. What did Sam learn about the different modes of transport during his journey?

4. Why are bicycles important for short trips and staying healthy?

5. How do cars, trains, ships, and airplanes each play a unique role in transportation?

6. What did Sam see from the airplane window that made him appreciate all the ways people travel?

7. How did transport help Sam explore new cities and cultures?

8. What is the role of transport in connecting people and places around the world?

9. What did Sam decide he wanted to do when he grew up?

10. How can you explore different modes of transport in your daily life?

Chapter 13 - Our Family

In the cozy town of Familiaville, families were the heart and soul of the community. Everyone in Familiaville knew the importance of family, and they cherished the time they spent together. The town was filled with laughter, love, and the warmth of family bonds.

A little girl named Sophie lived in Familiaville with her parents, grandparents, and younger brother. Her family was very close, and they did everything together—cooking meals, playing games, going on picnics, and sharing stories. Sophie's favorite time of day was when they all gathered around the dinner table, sharing stories about their day.

One day, Sophie's teacher gave her class an assignment: to create a family tree. Sophie was excited to learn more about her family's history, so she asked her parents and grandparents to help her with the project. They spent the evening looking through old photo albums, talking about their ancestors, and writing down the names of relatives.

As they worked on the family tree, Sophie learned about her great-grandparents who had moved to Familiaville many years ago, her uncle who was a firefighter, and her aunt who loved to paint. She discovered that her family had a rich history filled with interesting stories and traditions.

Sophie's grandparents shared stories about their childhood, how they met, and the challenges they faced growing up. Her parents talked about how they started their own family and the values they wanted to pass on to Sophie and her brother. Sophie realized that her family was like a tree with deep roots and strong branches, growing and thriving because of the love and care they all shared.

The project made Sophie appreciate her family even more. She understood that her family was her support system, always there for her in good times and bad. She felt proud to be part of such a wonderful family, and she knew that she would continue to cherish and strengthen these bonds as she grew older.

Moral: Family is the foundation of our lives, providing love, support, and a sense of belonging.

Activities:

1. Create a Family Tree:

Materials Needed: Paper, markers, and stickers.

Instructions: Have the children create their own family tree, including their parents, siblings, grandparents, and other relatives. Encourage them to ask their family members about their history and share stories.

2. Family Picture Frame:

Materials Needed: Craft sticks, glue, markers, and decorations (e.g., stickers, buttons).

Instructions: Help the children make a picture frame using craft sticks. Let them decorate the frame and add a family photo to it. Discuss the importance of keeping memories of family close.

Subjective Questions:

1. What did Sophie learn while creating her family tree?

2. How did Sophie's family help her with the project?

3. What stories did Sophie's grandparents share about their childhood?

4. How did Sophie's parents describe the values they wanted to pass on to their children?

5. Why is it important to know about your family's history and traditions?

6. How does a family provide support and love?

7. What did Sophie realize about the role of family in her life?

8. How can you strengthen the bonds with your own family?

9. What traditions does your family have that you cherish?

10. Why is it important to spend time with your family and share stories together?

Notes and Additional Resources

Printed at Repro India Ltd.